Extension Textbook

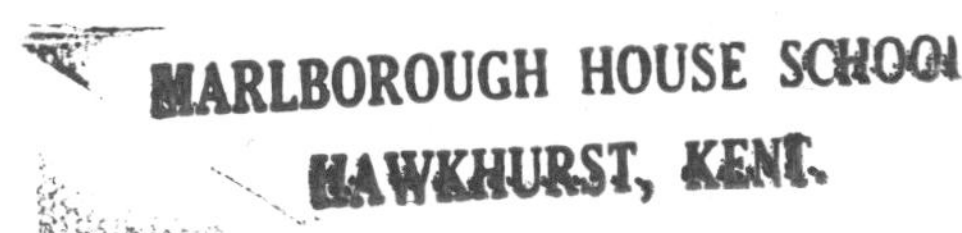

Heinemann Educational Publishers
Halley Court, Jordan Hill, Oxford OX2 8EJ
a division of Reed Educational and Professional Publishing Ltd

Heinemann is a registered trademark of Reed Educational and Professional Publishing Ltd

Writing team
John T Blair
Percy W Farren
Myra A Pearson
John W Thayers
David K Thomson

First Published 2001

06 05 04 03
10 9 8 7 6 5 4

ISBN 0 435 17424 X

Typeset and layout by Mandy Emery.
Designed by Gecko Ltd.
Illustrated by Teri Gower, David Till and Debbie Clark.
Cover Illustation by John Haslam.
Printed and bound by Edelvives, Zaragoza.

Contents

1 Copy and complete each sequence.

(a)	10000	20000	30000			
(b)		80000		60000	50000	
(c)	300000	400000			700000	
(d)				600000	500000	400000
(e)	1000000	2000000			5000000	
(f)				5000000	4000000	3000000

2 Write the number

- ten thousand more than

(a) 17 000 **(b)** 50 000 **(c)** 90 000

- ten thousand less than

(d) 70 000 **(e)** 20 000 **(f)** 11 000

- one hundred thousand more than

(g) 300 000 **(h)** 13 000 **(i)** 900 000

- one hundred thousand less than

(j) 800 000 **(k)** 400 000 **(l)** 200 000.

3 Write the number

- one million more than **(a)** 5 000 000 **(b)** 10 000 000
- one million less than **(c)** 2 000 000 **(d)** 10 000 000.

1 Copy and complete each sequence.

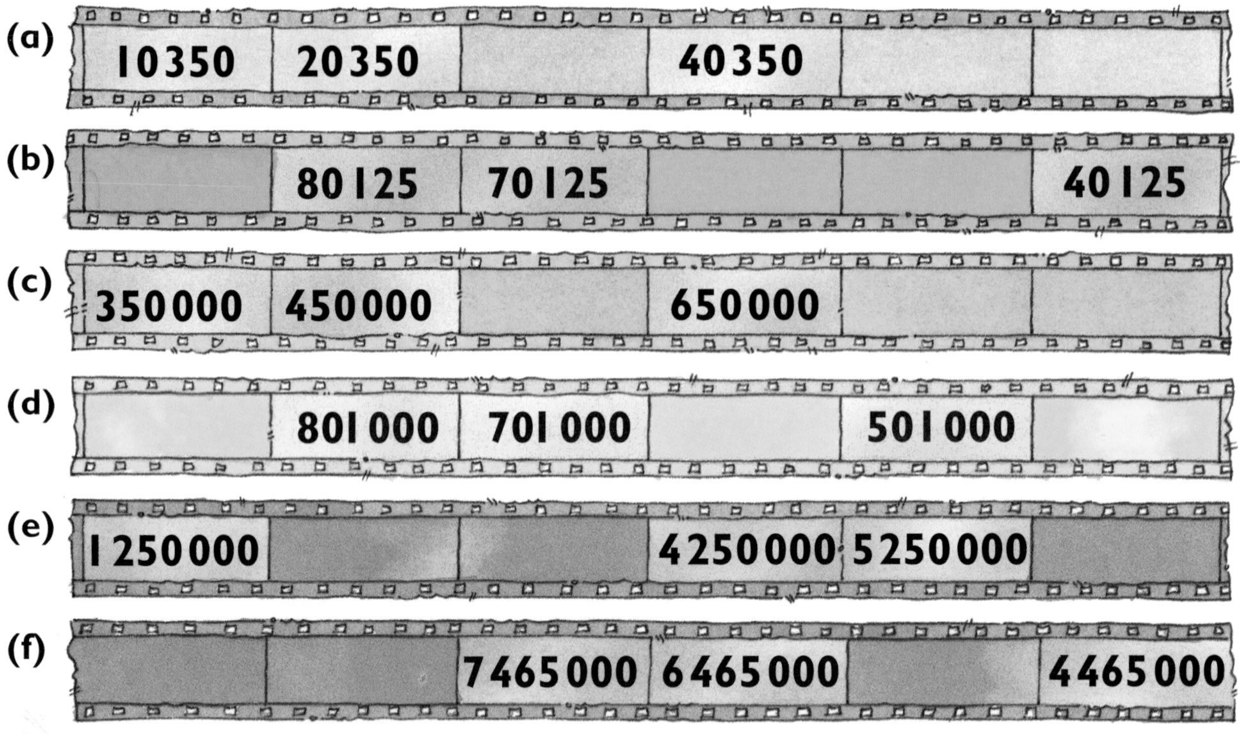

2 Write the larger number.

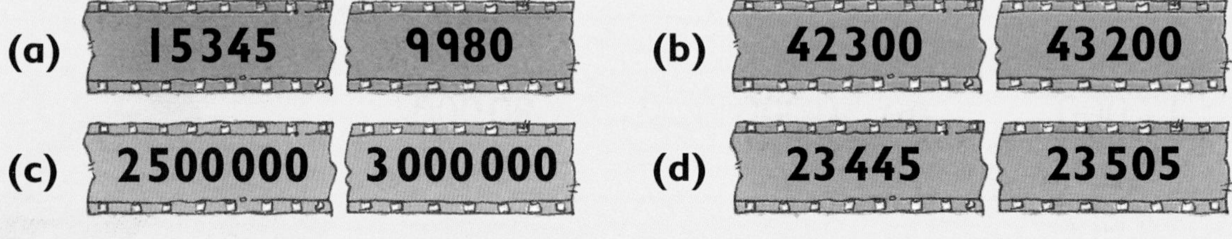

3 Write the smaller number.

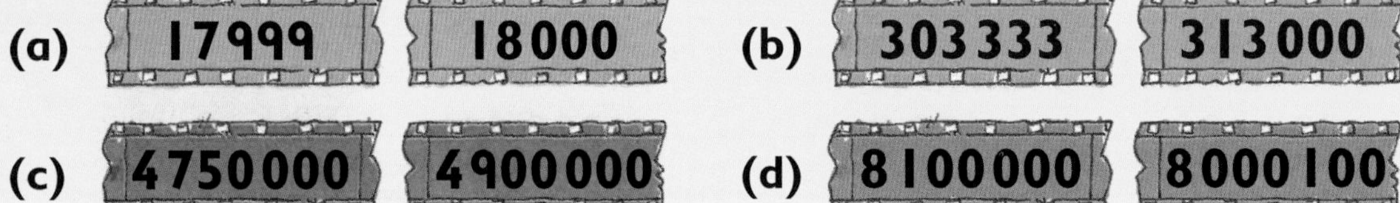

4 Write in order, starting with the largest number.

TREE SURVEY	
Walden Wood	
Type	Number
Oak	258
Elm	79
Lime	7
Ash	343

TREE SURVEY	
Fern Forest	
Type	Number
Pine	476
Spruce	9
Larch	195
Fir	64

1 What was the total number of trees found in

(a) Walden Wood **(b)** Fern Forest?

2 **(a)** 16 + 105 + 7 + 82 **(b)** 98 + 6 + 9 + 857

(c) 437 + 219 + 4 + 65 **(d)** 613 + 2 + 199 + 73

(e) 15 + 49 + 326 + 60 + 4 **(f)** 121 + 36 + 380 + 22 + 10

3 In Conway Copse, the survey found

- eight Plane trees
- two hundred and thirty Birch trees
- forty-seven Poplar trees
- three Whitebeam trees
- three hundred and sixty-two Beech trees.

How many trees altogether were found in Conway Copse?

I drive 75 miles every **day**.

I run 43 miles a week.

I walk 36 miles a week.

I cycle 28 miles each week.

Rob **Jan** **Martha** **Alan**

1 How many miles does

(a) Jan run in 6 weeks

(b) Alan cycle in 8 weeks

(c) Martha walk in 9 weeks

(d) Rob drive in 1 **week**?

2 Who travels more miles, Alan in 9 weeks or Martha in 7 weeks?

3 **(a)** 6 x 81 **(b)** 8 x 17 **(c)** 7 x 62 **(d)** 9 x 54
(e) 7 x 39 **(f)** 6 x 55 **(g)** 9 x 26 **(h)** 8 x 48

4

For the next eight weeks I will run an extra nine miles each week.

How many miles will Jan run in these eight weeks?

1 These sweets are packed in boxes of 8.

How many boxes are needed?

(a) 104 orange crunch **(b)** 152 strawberry cream **(c)** 136 lime lumps

2 These chocolates are packed in boxes of 6.

How many boxes are needed for

(a) 78 mint **(b)** 114 coffee **(c)** 102 coconut ?

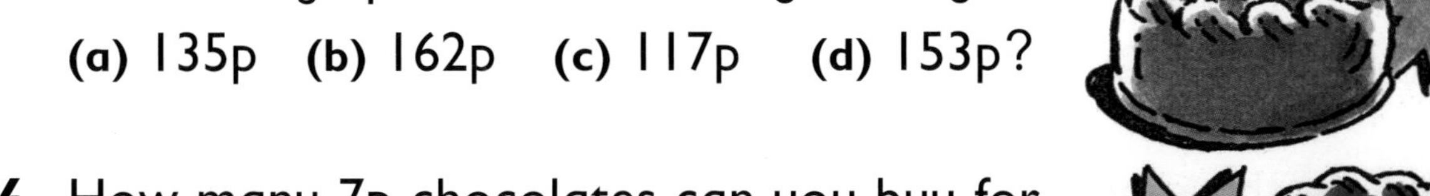

3 How many 9p chocolates can you buy for

(a) 135p **(b)** 162p **(c)** 117p **(d)** 153p?

4 How many 7p chocolates can you buy for

(a) 91p **(b)** 126p **(c)** 112p **(d)** 133p?

5 **(a)** Who bought more sweets?

(b) How many more?

E6

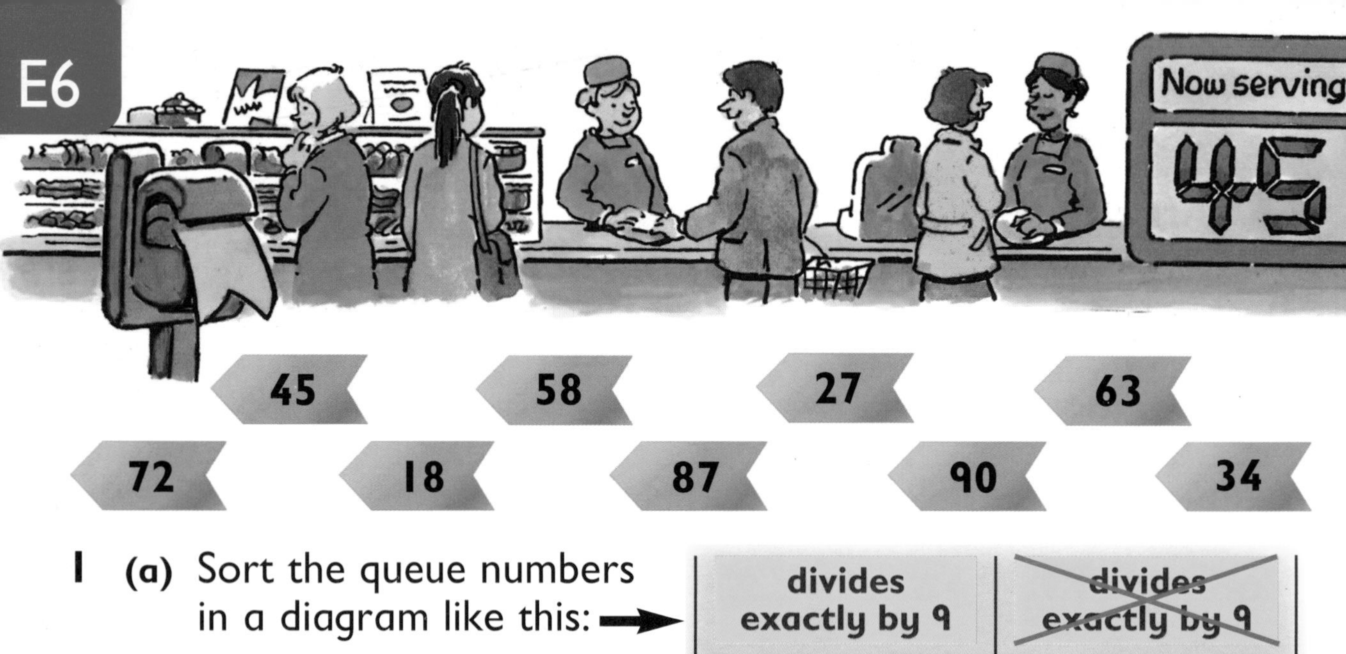

1 **(a)** Sort the queue numbers in a diagram like this: →

divides exactly by 9	~~divides exactly by 9~~

(b) The sum of the digits of 45 is 4+5 = **9**.

Copy and complete.

Numbers which divide exactly by 9	9	18	27	36	45	54	63	72	81	90
Sum of the digits										

(c) What do you notice?

2 27 14 22 30 6 25 18

(a) Sort these queue numbers in a diagram like this: →

divides exactly by 3	~~divides exactly by 3~~

(b) Copy and complete.

Numbers which divide exactly by 3	12	15	24	36	48	57	69	87	93	99
Sum of the digits										

(c) What do you notice about the sum of these digits?

3 **Without dividing**, find the numbers from 40 to 60 which

(a) divide exactly by 3 **(b)** do **not** divide exactly by 9.

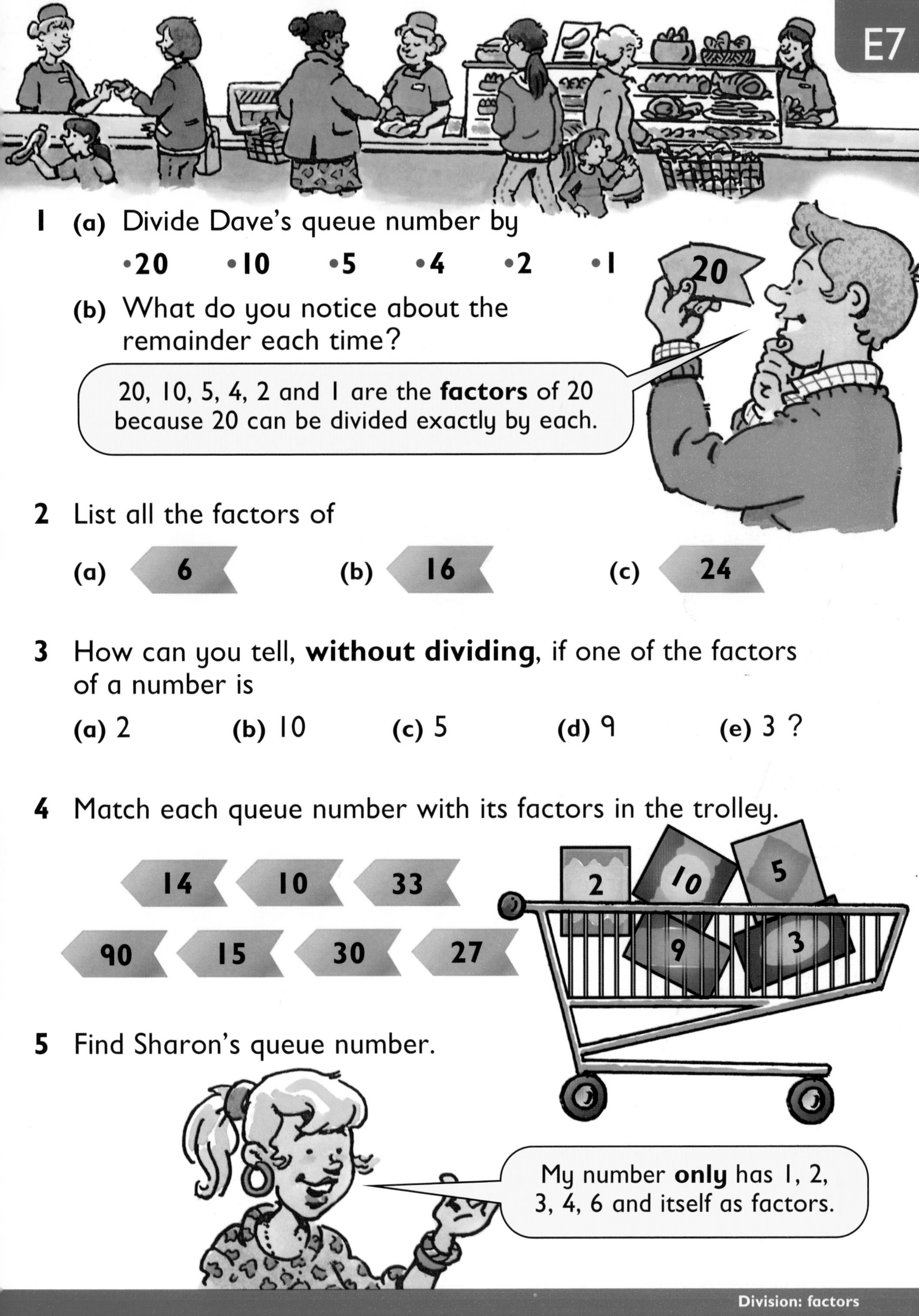

1 **(a)** Divide Dave's queue number by

•**20** •**10** •**5** •**4** •**2** •**1**

(b) What do you notice about the remainder each time?

2 List all the factors of

(a) 6 **(b)** 16 **(c)** 24

3 How can you tell, **without dividing**, if one of the factors of a number is

(a) 2 **(b)** 10 **(c)** 5 **(d)** 9 **(e)** 3 ?

4 Match each queue number with its factors in the trolley.

14 10 33 90 15 30 27

5 Find Sharon's queue number.

1 **(a)** $\frac{1}{6}$ of 24

(b) $\frac{1}{6}$ of 42

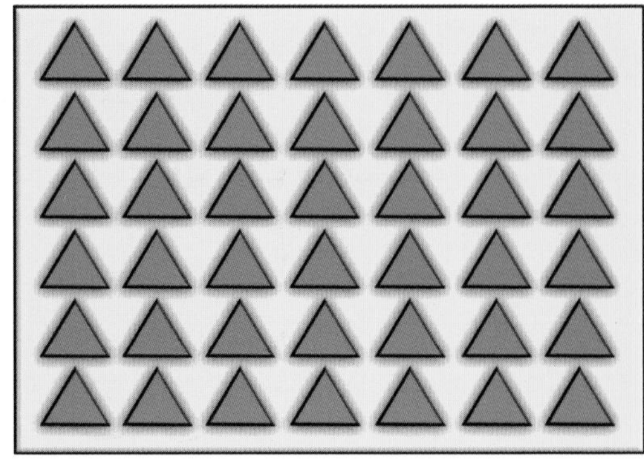

(c) $\frac{1}{6}$ of 30

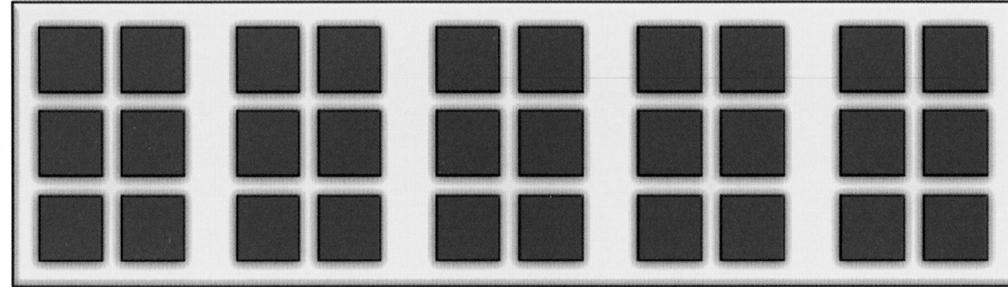

2 **(a)** $\frac{1}{6}$ of 18 **(b)** one sixth of 60 **(c)** $\frac{1}{6}$ of 48

(d) $\frac{1}{6}$ of 36 **(e)** one sixth of 6 **(f)** $\frac{1}{6}$ of 54

3 Alex has 42 shape stickers.
Half are circles, three are hexagons and one sixth of the rest are octagons.

How many stickers are octagons?

4

> One sixth of my stickers are pentagons.
> I have two pentagon stickers.

How many stickers has Kelly altogether?

1 **(a)** $\frac{1}{8}$ of 40 **(b)** $\frac{1}{8}$ of 24

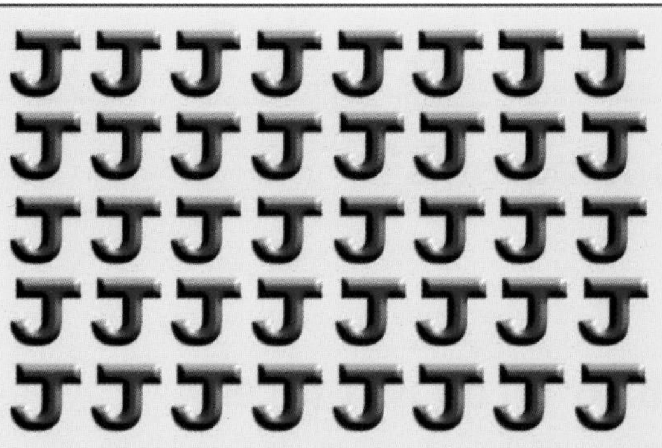

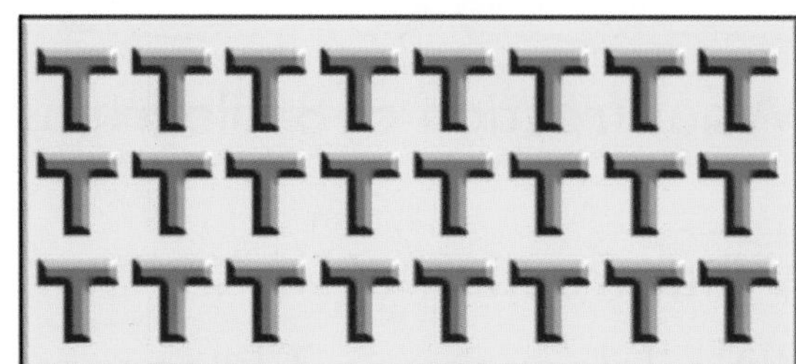

(c) $\frac{1}{8}$ of 32

AAAAAAAA AAAAAAAA
AAAAAAAA AAAAAAAA

2 **(a)** $\frac{1}{8}$ of 64 **(b)** $\frac{1}{8}$ of 72 **(c)** one eighth of 16

(d) $\frac{1}{8}$ of 80 **(e)** one eighth of 8 **(f)** $\frac{1}{8}$ of 56

3 Rachael has 48 badges.
One eighth are plastic and
the rest are metal.

How many of her badges are metal?

4 Class 4 wins 60 merit badges.

- Half are red.
- One tenth are green.
- One eighth of the rest are yellow.

How many badges are yellow?

1 

What fraction of 8 kilograms is 5 kilograms?

2 What fraction of

(a) 5 kilograms is 3 kilograms (b) 6 litres is 5 litres

(c) 3 litres is 2 litres (d) 4 metres is 3 metres

(e) £10 is £3 (f) £10 is £8?

3

1 metre

25cm

10cm

What fraction of one metre is

(a) 25 cm (b) 75 cm (c) 50 cm (d) 10 cm (e) 30 cm (f) 90 cm?

4 What fraction of **one pound** is

(a) 10 pence (b) 40 pence (c) 70 pence (d) 20 pence?

5 What fraction of the larger shape is the smaller shape?

(a)

(c)

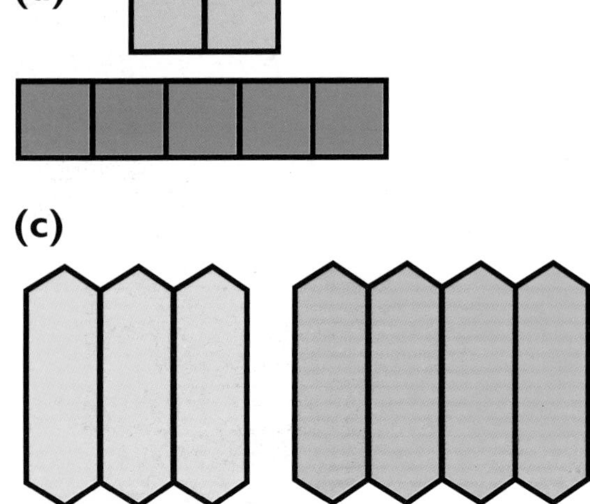

(b)

(d)

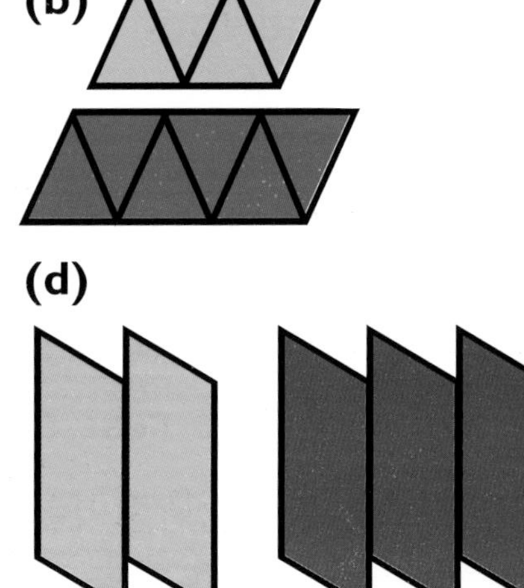

School shop-Amounts spent

	Mon	Tue	Wed	Thu	Fri
Class 3	£2·43	£2·56	£2·63	£1·78	£2·27
Class 4	£2·72	£1·52	£3·27	£2·65	£3·49
Class 5	£3·89	£2·21	£3·22	£1·82	£3·74

1 Find the total amount spent by Class 3 on Monday and Tuesday.

2 How much was spent altogether by Class 4 on Wednesday and Thursday?

3 Which class spent the greatest total amount on Thursday and Friday?

4 How much more was spent by Class 5 than by Class 4 on Monday?

5 How much less was spent by Class 3 on Thursday than by
(a) Class 5 **(b)** Class 4?

6 Find the difference between the amounts spent by Class 3 and Class 5 on Wednesday.

7 On which **day** was the total amount spent
(a) greatest **(b)** least?

1 Which machine has the highest total score?

(a)
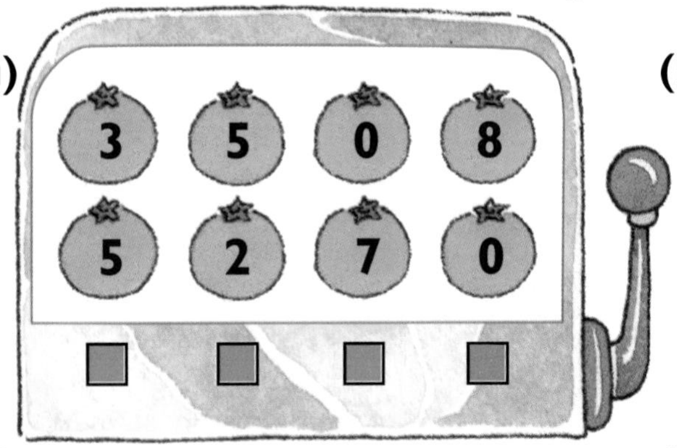
3 5 0 8
5 2 7 0

(b)
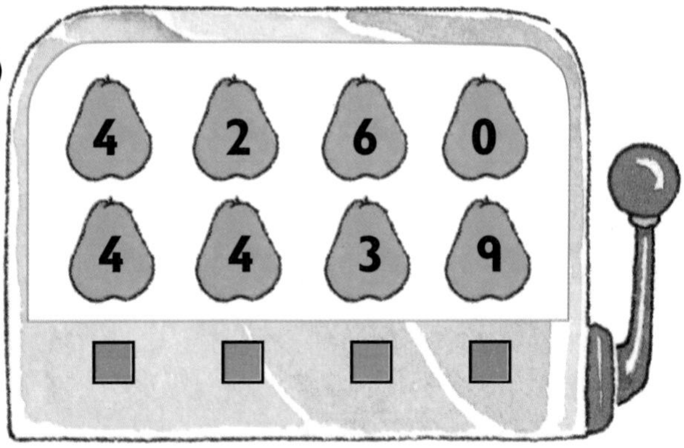
4 2 6 0
4 4 3 9

(c)
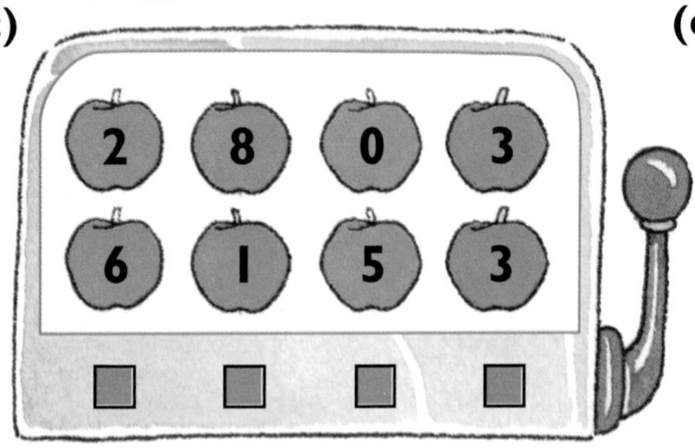
2 8 0 3
6 1 5 3

(d)
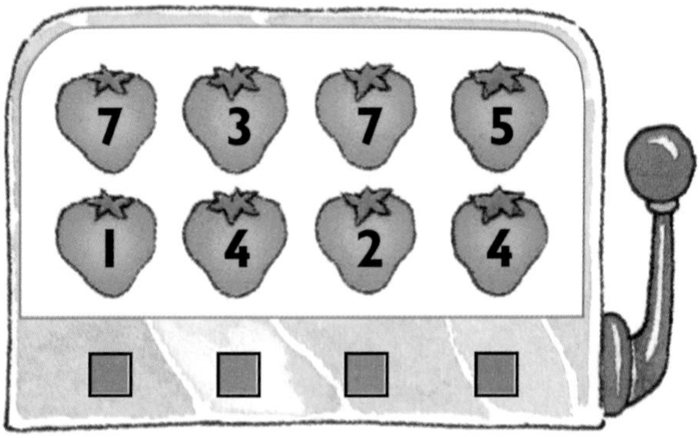
7 3 7 5
1 4 2 4

2 **(a)** 1530 + 6069 **(b)** 2765 + 6104 **(c)** 3017 + 4280
(d) 3243 + 3216 **(e)** 2254 + 1012 **(f)** 5555 + 1342

3 Which pairs of scores add to give the same totals?

(a)
4 3 2 4

(b)
3 2 0 4

(c)
3 1 2 4

(d)
4 2 4 4

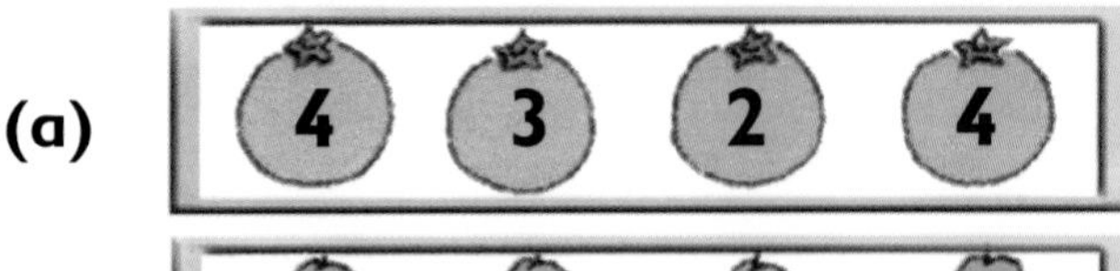

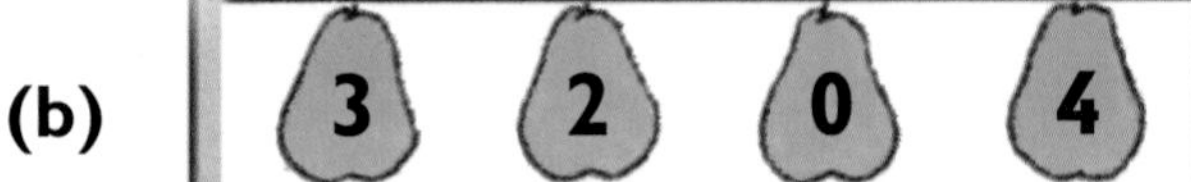

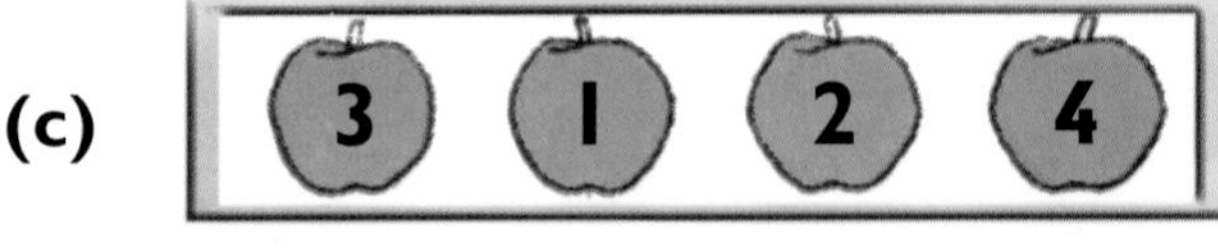

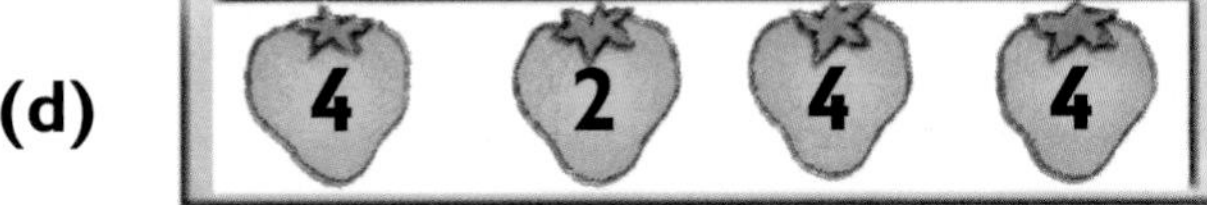

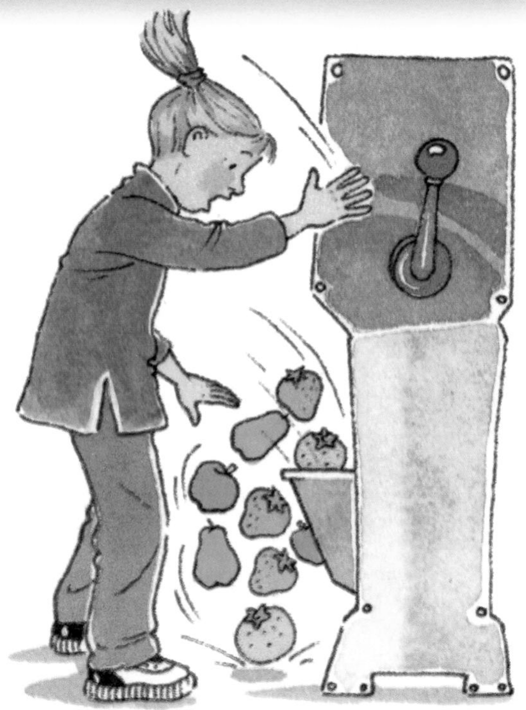

GREAT DEALS ON USED CARS

BEST PRICES

Honda
£5473

1 Find the difference in price between these cars:

(a) Rover and Vauxhall

(b) Ford and Fiat

(c) Saab and Mazda

(d) Chrysler and Honda.

2 **(a)** Which car costs £2112 less then the Toyota?

(b) Which car costs £5110 less than the Mazda?

3 **(a)** 8747 – 5624 **(b)** 7696 – 3253 **(c)** 6434 – 2332

(d) 3883 – 1512 **(e)** 8482 – 4431 **(f)** 7361 – 6320

1 Teri, George and Shareen start a computer money game with £1234 each.

How much does each player have at the end of the game?

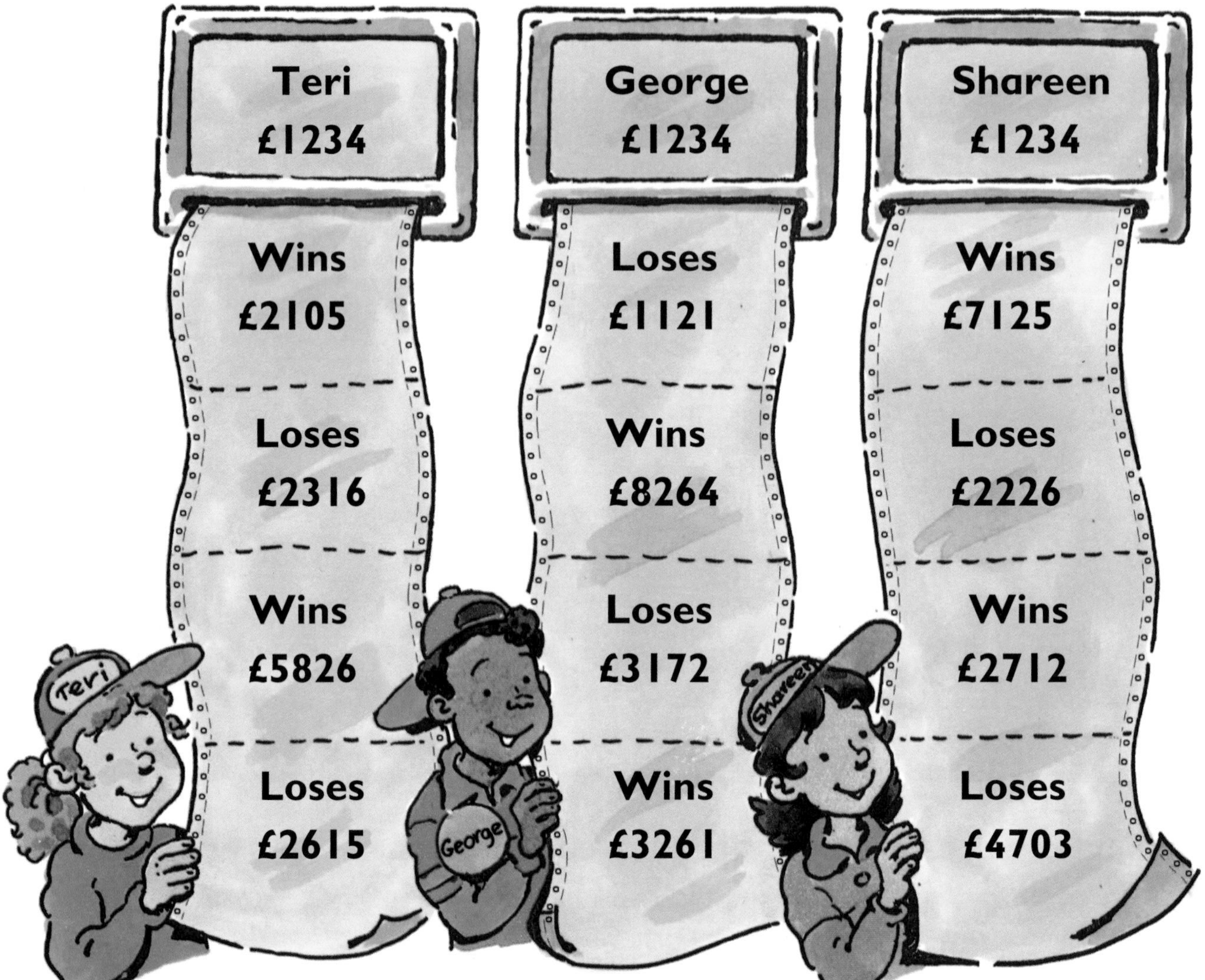
Teri
£1234
Wins
£2105
Loses
£2316
Wins
£5826
Loses
£2615
George
£1234
Loses
£1121
Wins
£8264
Loses
£3172
Wins
£3261
Shareen
£1234
Wins
£7125
Loses
£2226
Wins
£2712
Loses
£4703
Teri
George
Shareen

2 What is the difference between the greatest and least amounts of money held by players at the end of the game?

3

£3978
£4163
£2315
£5726

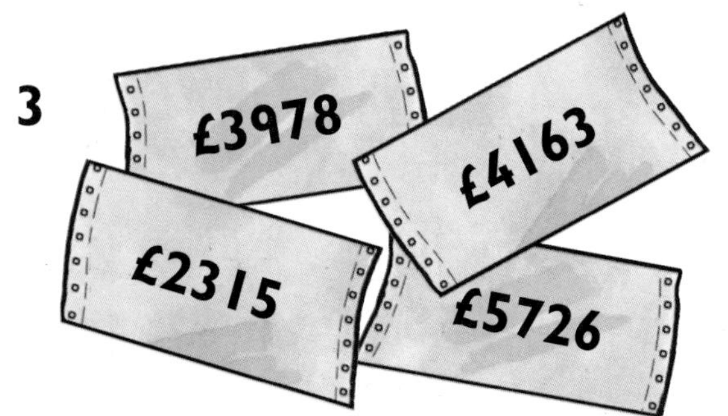

Which two amounts have

(a) a total of £6478

(b) a difference of £1663?

1 Write the time

(a) 17 minutes after **(b)** 23 minutes after **(c)** 36 minutes after

(d) 8 minutes before **(e)** 24 minutes before **(f)** 42 minutes before.

2 How many minutes are there between each of these times?

(a)

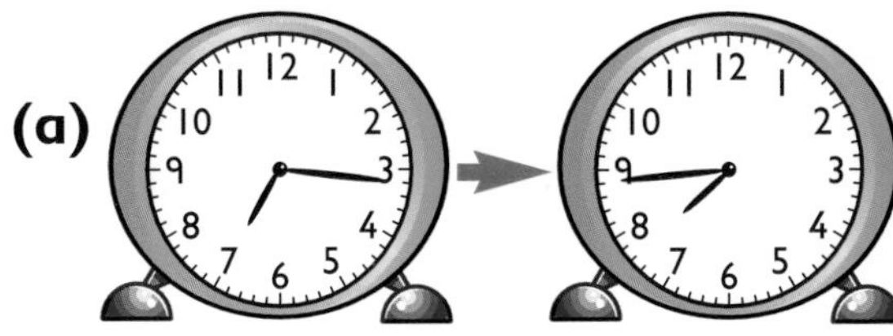

(b)

3 Hannah started a puzzle at 3.13pm. She took 38 minutes to complete it.

At what time did she finish?

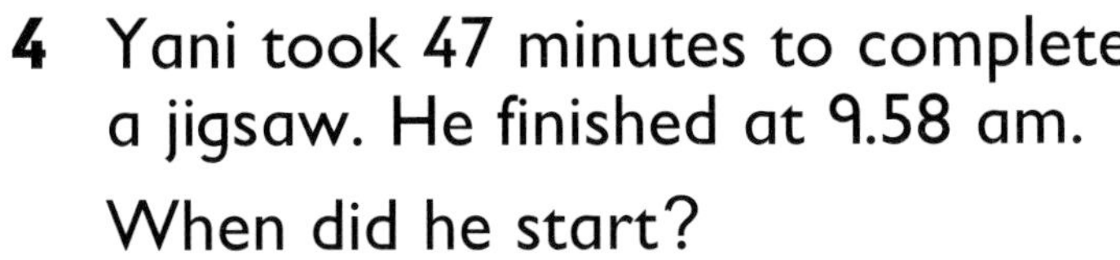

4 Yani took 47 minutes to complete a jigsaw. He finished at 9.58 am.

When did he start?

5 Eric and Petra started playing at 12.11 pm and finished at 12.50 pm.

For how long did they play?

1000 metres = 1 kilometre (km)

1 Write these distances in **kilometres**.

(a) 2000 m **(b)** 10 000 m **(c)** 25 000 m **(d)** 8500 m

2 Write in **metres**.

(a) 3 km **(b)** 15 km **(c)** $7\frac{1}{2}$ km **(d)** 10 km 500 m

3 List in order, starting with the shortest distance.

5 km	**5 km 400 m**	**$5\frac{1}{2}$ km**	**500 m**	**5900 m**

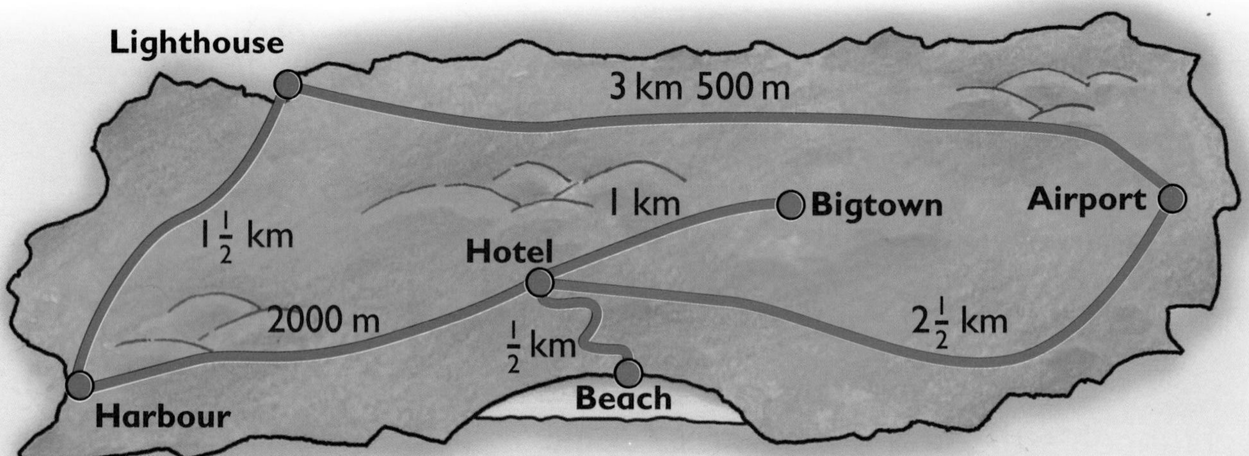

4 Write in metres the shortest distance from

(a) Bigtown to the Harbour **(b)** the Airport to the Harbour

(c) the Beach to the Lighthouse.

5 John lives halfway between the Harbour and the Hotel.

How far is it from his home to the Beach?

6 Use a trundle wheel to mark out a length of 50 m.

(a) Find how long it takes you to walk 100 m.

(b) **Estimate** how long it would take you to walk 1 km.

1 centimetre = 10 millimetres (mm)

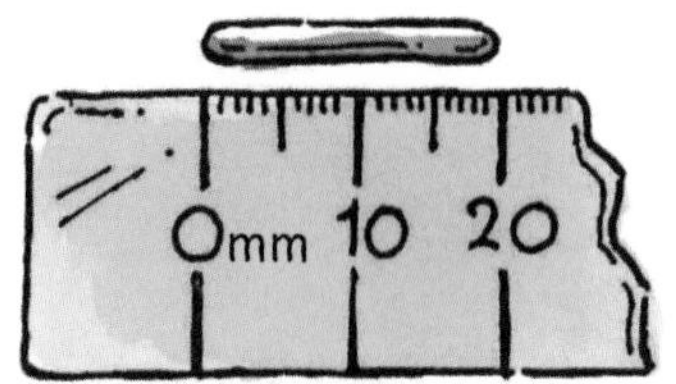

The worm is... about 2 centimetres long...about 20 millimetres long.

1 Write in **millimetres**.

(a) 8 cm **(b)** 10 cm **(c)** $13\frac{1}{2}$ cm **(d)** 20 cm **(e)** $25\frac{1}{2}$ cm

2 Write in **centimetres**.

(a) 40 mm **(b)** 85 mm **(c)** 210 mm **(d)** 170 mm **(e)** 105 mm

3 Find the length of each insect

- in centimetres
- in millimetres.

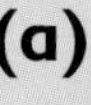

(a)

(b)

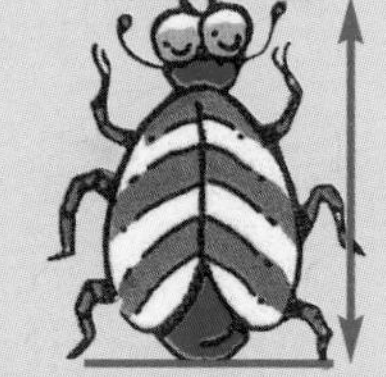

(c)

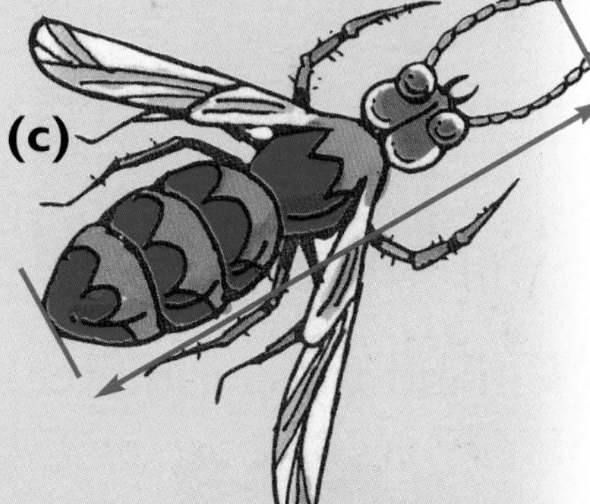

(d)

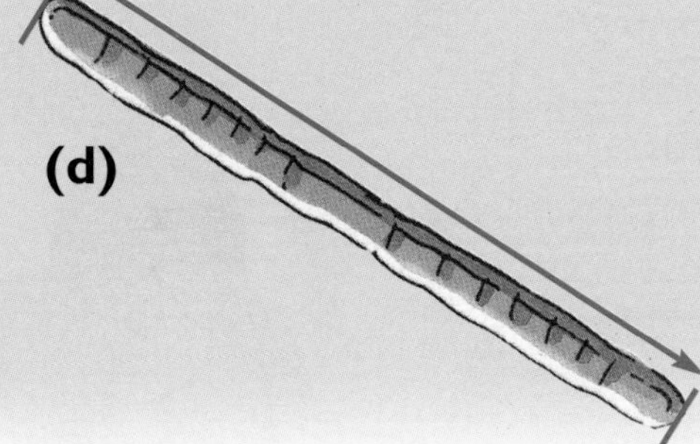

(e)

(f)

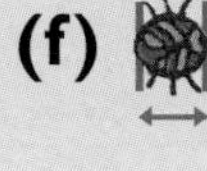

4 List these lengths in order, starting with the longest.

10 mm	$8\frac{1}{2}$ cm	2 cm	55 mm	30 mm	7 cm

1 Find the perimeter of each shape by measuring.

(a)

(b)

2 Find the perimeter of this shape.

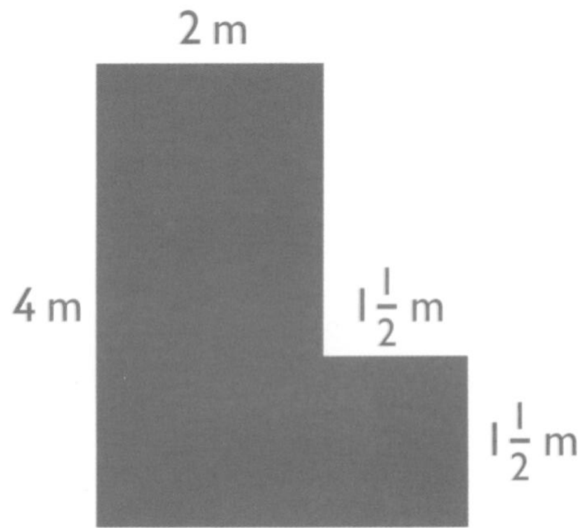

3 What is the width of this rectangle?

20 m　　Perimeter = 96 m　　20 m

4 Find the perimeter of each child's shape.

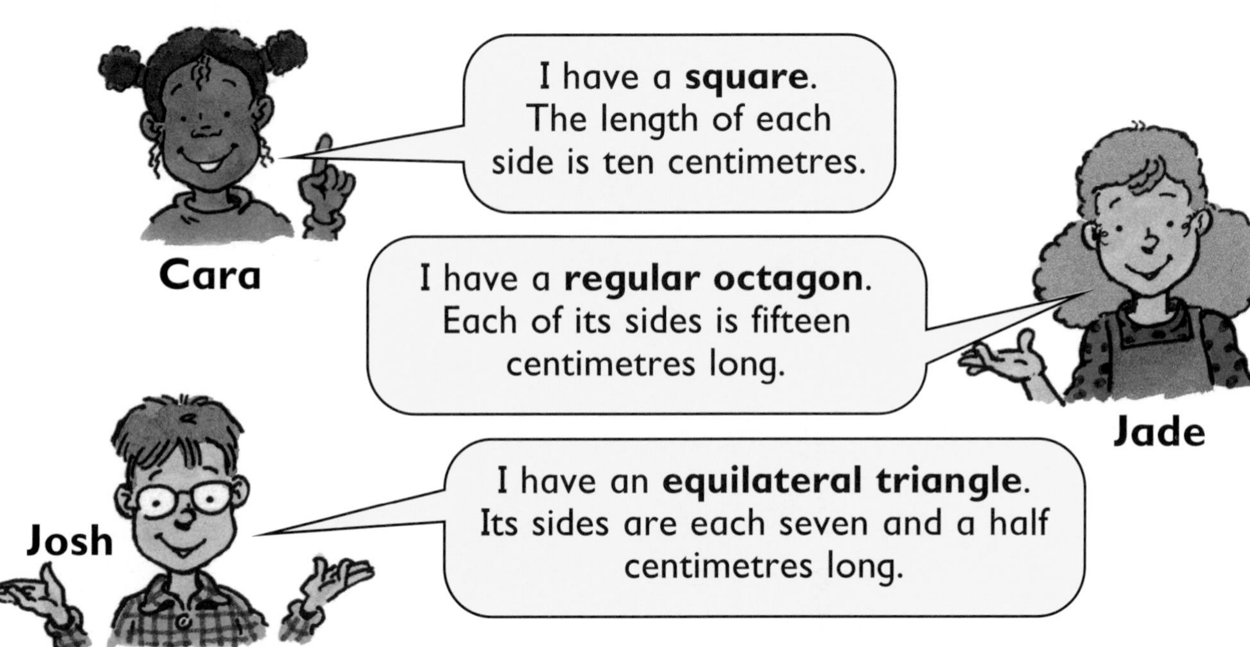

1 Write the weight of each item to the

(a) nearest 100 g **(b)** nearest 100 g **(c)** nearest 50 g

(d) nearest 50 g **(e)** nearest 20 g **(f)** nearest 10 g.

2 Work as a group.

Find the weight of items like these:

1 Who has given the best answer?

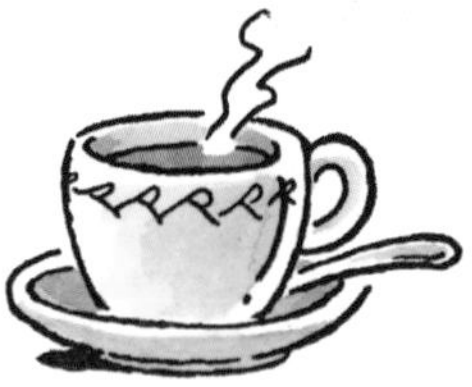

(a) The amount of tea in a cup…

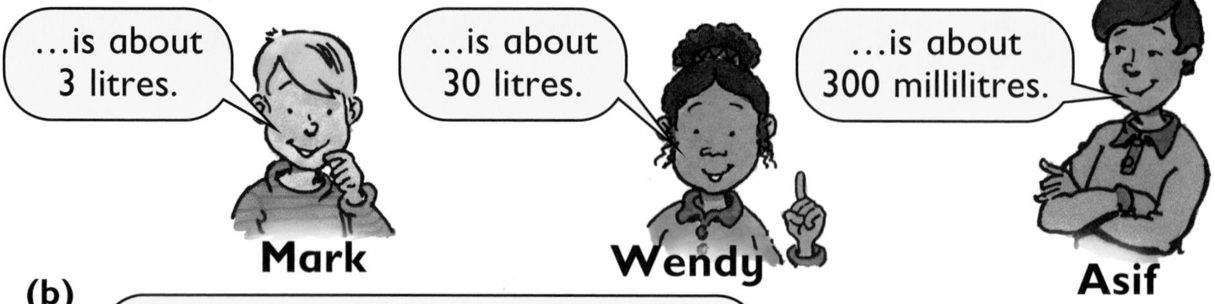

…is about
3 litres.
Mark
…is about
30 litres.
Wendy
…is about
300 millilitres.
Asif

(b) The **weight** of one litre of juice…

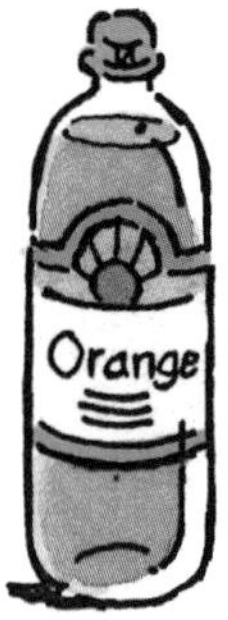

Orange

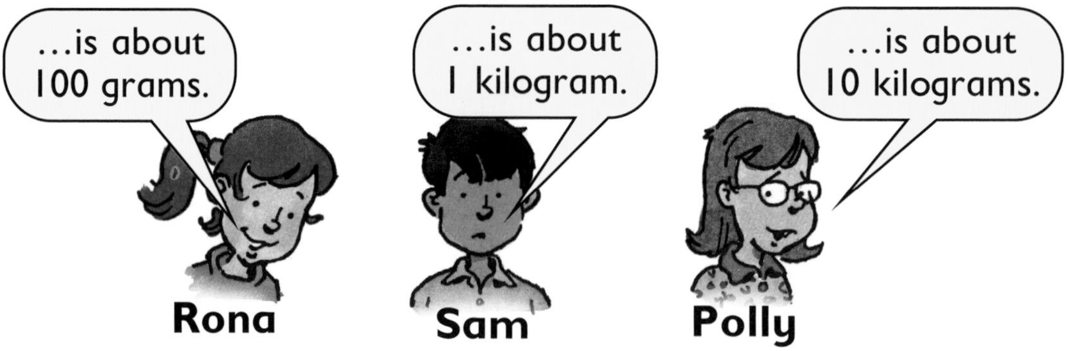

…is about
100 grams.
Rona
…is about
1 kilogram.
Sam
…is about
10 kilograms.
Polly

2 Use one of these labels to complete each sentence correctly.

1 km	20 cm	50 g	4 ℓ	200 m
400 ml	$4\ cm^2$	5 kg		20 mm
$40\ cm^2$	2 mm	50 cm		40 ℓ

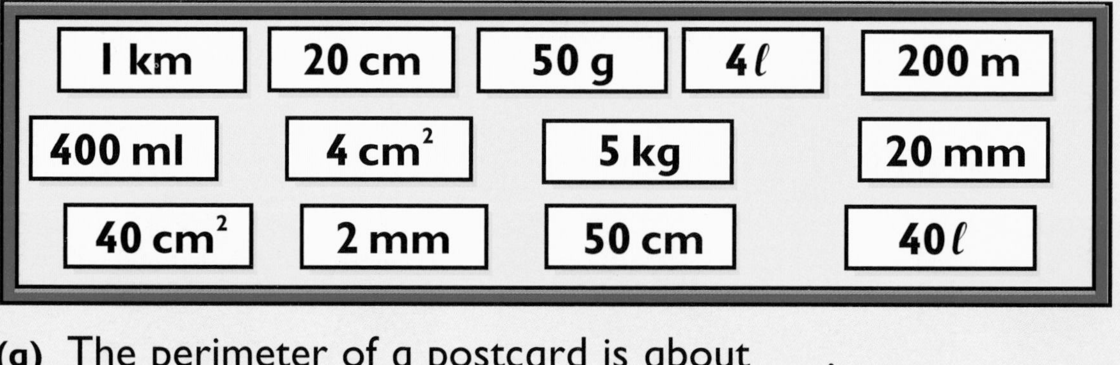

(a) The perimeter of a postcard is about ___.

(b) The amount of water in a goldfish bowl is about ___.

(c) The width of a one pound coin is about ___.

(d) The area of a playing card is about ___.

(e) The length of ten football pitches is about ___.

(f) The weight of a calculator is about ___.

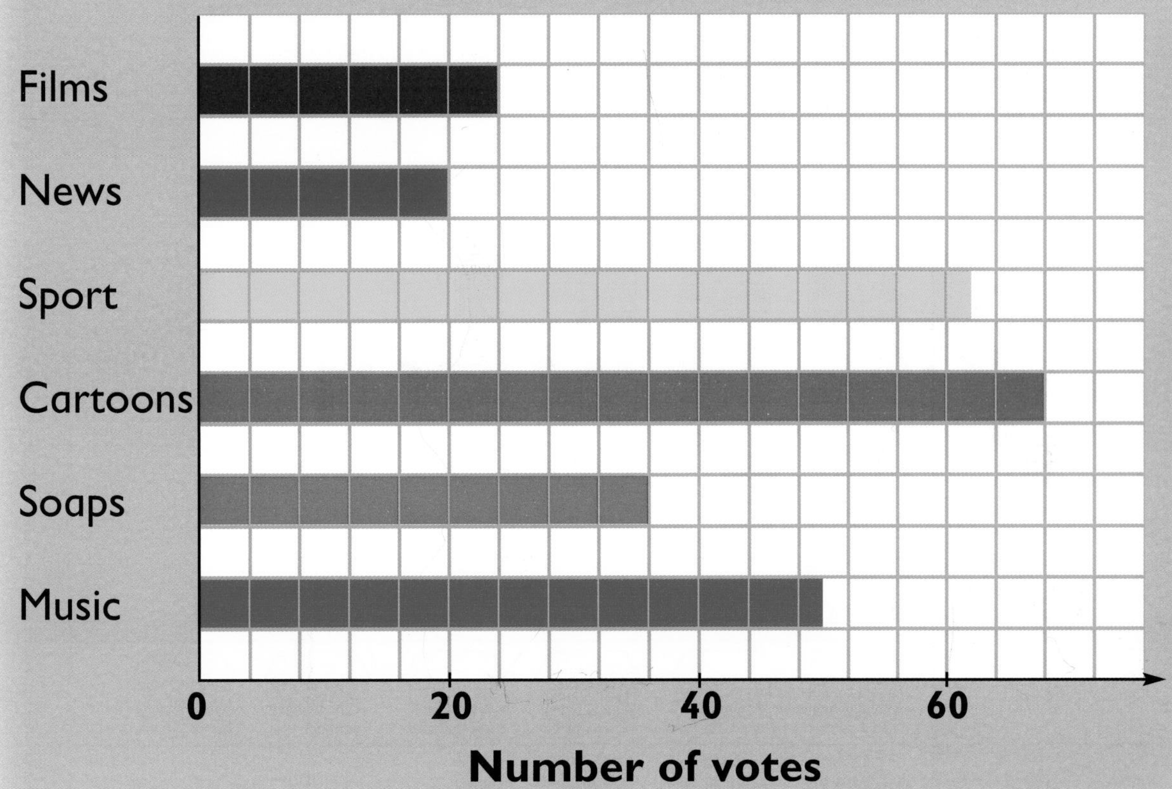

1 Which type of programme was
(a) most popular (b) least popular?

2 How many more children voted for
(a) Cartoons than Films (b) Music than Soaps?

3 Which two types of programme had a total of sixty votes?

4 How many children voted altogether?

5 Which two types of programme were voted for by half of the children?

Each represents 5 houses.

represents **between** 5 and 10 houses.

Houses on Glenview Estate

Street	
Grove	
Avenue	
Close	
Lane	
Terrace	
Crescent	

1 How many houses are in

(a) Glenview *Lane* **(b)** Glenview *Terrace*?

2 Which street has **between**

(a) 5 and 10 houses **(b)** 20 and 25 houses?

3 Which two streets have **about** the same number of houses?

4 Which streets have

(a) fewer than 15 houses **(b)** more than 20 houses?

5 Draw what the graph would show for Glenview *Wynd* which has 27 houses.